HOME SANITATION & WATER CONDITIONING REPAIRS

by Larry Elrod

THEODORE AUDEL & CO.

a division of

HOWARD W. SAMS & CO., INC.

4300 West 62nd Street
Indianapolis, Indiana 46268

FIRST EDITION

FIRST PRINTING—1974

International Standard Book Number: 0-672-23803-9

Contents

SECTION

1

Introduction

Regardless of your background, upbringing, or status in life, at one time or another, all people are temporarily humbled to the same level when something goes wrong with the plumbing. The seriousness of the problem can range from a pesky leak at a faucet or around a fixture to a toilet that won't flush. Friends and neighbors are very helpful in times of need and times of crises; just so these times of need don't happen too often or last too long when they do. Certainly one of these crises is when you must use your neighbor's bathroom for a week because of a plumbing problem.

The minor plumbing or related maintenance repairs around the household are far less disrupting to you and your neighbor's lifestyle. However, the potential expense of these repairs is what concerns most breadwinners who are trying to balance the monthly budget. There are some few people who in a social-impressing atmosphere among friends drop statements like, "my attorney," "we are good friends of the judge," or "my plumber." These people usually don't have a need for this book because they simply get on the phone and call a local plumbing shop or mechanical contractor that does repair work and within a few days a truck with all kinds of paraphernalia rolls up the drive and their worries are over as quickly as the personal check is signed. For the rest of us, *and* some of them, trying to fix plumbing problems ourselves means saving some money. That is what this book is all about.

This book is a companion to a book that covers almost all simple plumbing repairs that any homeowner may encounter with his faucets and drains. Herein is found the additional information for the care and servicing of the less frequent home plumbing problems not concerned with faucets, drains, and toilets.

It will help you to do those minor plumbing repairs and understand some elementary functions and "how things work." It will also provide you with the information to *think* about your plumbing care. Often the difference between a family who has one plumbing problem after another, and a family who hasn't had that first problem in six to eight years, is attitude and knowledge.

Many states and cities require licensed plumbers to do plumbing installation and repairs in commercial buildings and public places. Most building or plumbing codes also require plumbers to install new domestic plumbing when houses are built. However, virtually all homeowners *can* do their own plumbing repairs and need not be a licensed plumber.

With a little effort at learning and some practical common sense you too can fix your plumbing. It isn't all that difficult. The neighborhood kid who has worked on cars for 10 years since he was 15 probably was self-taught. If he does it regularly, he probably is pretty fair at his repair work. Regardless of how good he gets he puts up with the "shade-tree mechanic" title because he knows how much money he is saving. You too can piddle around with plumbing at your house and get pretty good at it. Who knows, you too may even earn that seemingly contemptuous title from your neighbors of "shade-tree plumber," but just think of the money you will be saving. The real and honest test of your ability and recognition as a do-it-yourself plumber comes when your friends and neighbors start coming to you for *free* advice.

Plumbing repairs are never inexpensive; neither are parts and supplies on small repair jobs that you do yourself. However, by doing it yourself, you can literally save a bundle of cash. The information and tips in this book are not meant to make you a journeyman plumber, but they should save you money and give you a limited working knowledge to help yourself.

Unless you have previous plumbing experience, repair work *beyond* the tips presented here should be approached with caution and with the phone number of the local plumber written on the wall by the phone.

TOOLS

Since you want to do the job right, a few specific tools will enable you to do so. Many of them you may already have. Good screwdrivers of the slot- and phillips-head type are required (Fig. 1). You should also have at least two wrenches. One is a 10″ pipe wrench (Stillson) and an adjustable end wrench as shown in Fig. 2. The pipe wrench has teeth in the

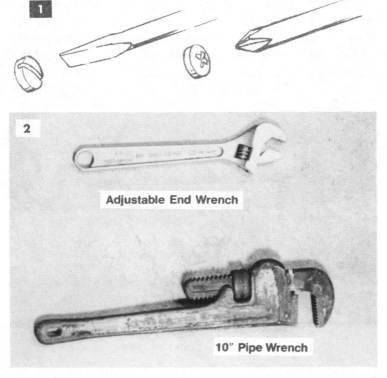

1

2

Adjustable End Wrench

10″ Pipe Wrench

jaws and this is the only common wrench you can buy that will hold or turn pipe. It will often be used as a "back-up" to hold something from turning while loosening or tightening something else. Pliers are useful everywhere. Standard pliers and the *Channellock®* style especially are good. The *Channellock®* are often used as a pipe wrench when heavy strains are not required to turn pipes (Fig. 3). Snakes are used to cut or bore through obstructions in drain lines, pipes, toilets, etc. A short *snake* or *closet auger* is fed to unclog most simple obstructions in household plumbing. The other kind of snake is a tightly wound and flexible instrument that can be bought in 25- and 50-foot lengths. A longer snake is necessary if you have an obstruction in a pipe between the house and sewer or septic tank. Snake costs are not high. But unless you have a recurring root-clogging or septic tank back-up problem you should rent or borrow exactly the snake you need (Fig. 4).

Common sense is another tool you should have. It will enable you to improvise all kind of gadgets to help you out. Some of which are wire hangers of any kind for supporting pipes and hoses strung overhead, and "coat-hanger" snakes.

Everyone should have a "plumbers friend," or a plunger (Fig. 5), along with a small assortment of other tools and

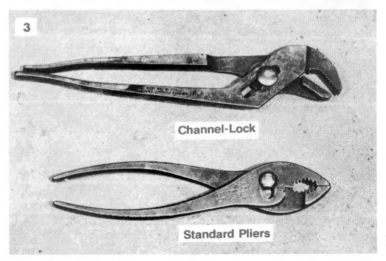

3

Channel-Lock

Standard Pliers

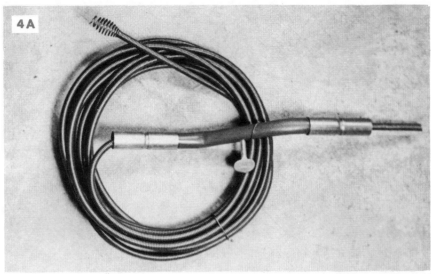

4 A

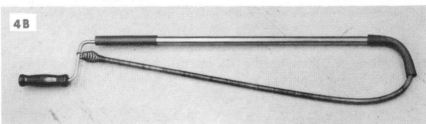

4 B

5

9

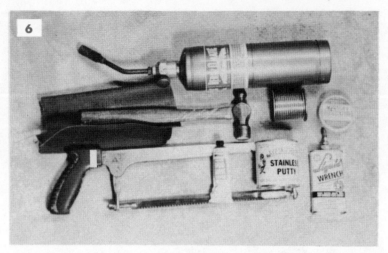

supplies like hammer, hacksaw, files, plastic tape, pipe dope of the *Blue Ribbon* or *John Sunshine* type, tape pipe dope, epoxies, tub caulking, solder, flux, small torch, and emery cloth (Fig. 6). Seldom will you need a torch unless a leaky copper pipe joint needs fixing or an entire faucet needs replacing. Otherwise a small torch is a handy tool to have for all kinds of heating jobs around the home.

SECTION

Water Supplies and Conditioning Equipment

Water conditioning equipment in the house ranges from an iron filter to possibly a water softener. Families that use city water get their water "conditioned" to some extent. It generally contains very little, if any, iron either ionized or in suspension, no silt or hydrogen sulfide taste, and is virtually guaranteed safe to drink. The probability of it being very hard is high. Many families prefer this hard city water that sometimes has a chlorine taste. Many do not and have a water softener installed or try to do it themselves.

Those families who depend on a well for water have a similar situation but often have more in their water to contend with. The other significant problem well users have is iron. Iron is often ionized (invisible) and/or in suspension. The suspended or "red" iron that you can see tends to make the color of the water brown or rust colored. This coloring in the water is bacteriologically harmless to its users. It does stain clothes, toilet bowls, and sinks.

The average home that has iron in the water can get rid of it with a typical water softener (Fig. 7). If you don't have a softener but would like one, call and visit two or three places that sell them and find out all about them. Have a salesman come out and test the water from each place you call or visit. Save the test results and compare them. This will give you some indication as to the relative "hardness" and iron content of the water. If all the tests result in similar

readings you should have a good idea of just what the condition of the water is. If any result is far higher than any of the others, you should question the credibility of that particular test.

The water may test hardness so low that a simple company-exchange type softener will do the job. This type of softener is a simple tank that is exchanged by the company on a regular basis. This is a very effective and desirable type. However, if you have iron or very hard water or have at least a family of four, a large higher-capacity brine recharging unit will be needed.

To install a brine recharging water softener, place it anywhere in the supply line between the well-pump pressure tank or city water supply valve in the basement or utility room and the first fixture that uses water (Fig. 8). The tools for this installation will need to be basically the ones listed in the front of this book. If copper pipe is used, you will need to know how to solder and "sweat" copper joints. This procedure is covered later in this book.

All softeners have a very simple hookup arrangement and you can do it. If you buy one from Sears or other like com-

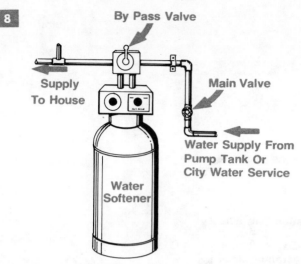

pany, detailed installation instructions will be included in the price. Very simply, all softeners have an inlet connection, an outlet connection, and a discharge connection.

Step 1. Decide where to put the softener.

Step 2. Make sure you have enough extra pipe, fittings, and the proper tools.

Step 3. A bypass valve will need to be installed. Work it a few times and determine which is the:

    A. Inlet from the supply,

    B. Outlet to the rest of the house,

    C. Outlet to the inlet of the softener.

    D. Inlet from the outlet of the softener.

Step 4. Install the bypass valve in the main supply line similar to that shown in Fig. 9.

Some water softeners have a built-in bypass valve. If so, omit steps 3. and 4.

Step 5. Connect the inlet opening of the softener to the water supply line (if the softener has a built in valve) or to the outlet of the bypass valve.

Step 6. Connect the outlet opening of the softener to the supply line going to the rest of the house or to the bypass valve.

Step 7. Check the bypass valve for proper working order. It should (a) completely direct water

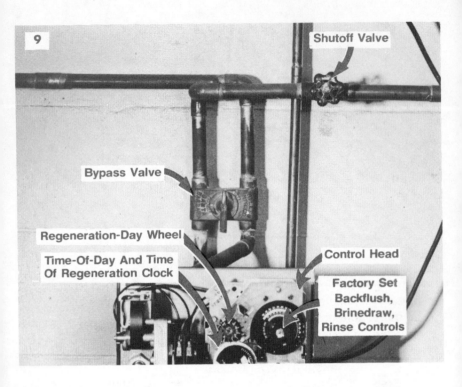

through the softener and (b) completely bypass
the softener as if it wasn't there.

Step 8. Connect a water/brine discharge line to the
discharge outlet of the softener and run this
line to the sewer through a floor drain or to a
sump pump.

Step 9. Check all joints for leaks.

Step 10. Plug the controls clock to an electrical outlet
and set the correct time on the clock. Also set
the clock for the desired regeneration time.
Generally all other times for brine draw, back
flushing, and rinsing are factory preset. If not,
each manufacturer will provide explicit instruc-
tion for doing this—follow them.

Step 11. Set the days on which regeneration is to take
place. How often regeneration takes place de-

pends on the total water usage of any family. A simple indicator is to observe if the water gets hard routinely. If so, regenerate more often. As a place to start, set the regeneration day to every fourth day or follow the manufacturer's instructions. Adjust regeneration days to suit soft-water needs.

Step 12. Put salt in the brine tank. The softener is ready to use.

If the water softener is not operating properly and you are renting it, call the company serviceman. It is their unit and responsibility to service it. If you bought it, then do the following:

Step 1. Check the salt level. Keep salt above the brine level in the salt storage tank.

Step 2. Try to manually regenerate.

Step 3. Check the clock to see if the power has been or is off. If so, reset the clock. Check (listen) to see if clock motor is running.

Step 4. Check the brine strainer and water-level float valve for proper operation (Fig. 10). If float

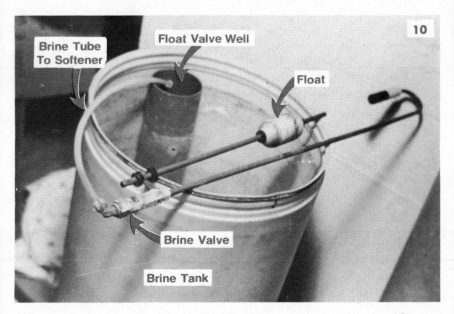

valve is bad, replace it. You may have to replace the entire brine-valve assembly.

Step 5. If *rock* salt has been used, clean out the bottom of the brine tank. Often very fine sand accumulates to a depth of several inches and impairs operation of the brine float valve and brine strainer. If salt *pellets* have been used, sand isn't likely to accumulate in the bottom of the brine tank since these pellets are clean pure salt. Rock salt has been washed but still has a lot of sand in it.

Step 6. Set to regenerate more often; at least every other day.

Step 7. If the unit is doing double duty as an iron filter and a softener, periodically mix a commercial iron remover, such as *Iron Out,* with the salt.

Step 8. If all else fails, throw the bypass valve to eliminate the filter from the line and call the company service man. If the unit has been in use for several years, the ion-exchange mineral in the mineral tank probably needs replacement.

IRON FILTER

If iron is a problem and a filter is needed, shop around much the same as you would for a softener. Get several tests then make a decision. You can install an iron filter in the supply line but it should be placed in before any softener that is in the line.

Simple cartridge filters can be used which have a replaceable-filter element much like the oil filter on a car. Usually these are for small homes where the amount of suspended iron and the water demand is small. After the element has been replaced by a new one over a hundred times, this chore becomes very troublesome. Unless these filters are used in the water supply line of a single appliance or sink they should not be installed.

Any serious effort to eliminate the iron from a household water supply calls for the rechargeable-tank type (Fig. 11).

The connections, discharge line and clock mechanism are hooked up exactly like those for a softener. Follow the installation steps provided by the manufacturer and/or those listed in this book under softeners.

Servicing an iron filter is simpler than a softener.

Step 1. Follow the manufacturer's recommendation for recharging the mineral with potassium permanganate at least every six weeks.

Step 2. Check the controls clock and back-flushing indicator for proper operation according to the manufacturer's specifications included with the unit.

Step 3. If the iron filter has been in operation for several years, the mineral or "green sand" may need replacing. If all else fails, call the company service department and have the mineral replaced. You can do it but it is a very messy job. Simply disconnect and take the mineral tank outside and lay it on its side. Remove the top and dig and flush out the mineral. Buy new sand from the manufacturer and put it in after the tank has been installed back in the water supply line.

Step 4. If this tank or any tank has a sweating problem in a damp basement, cover it with an old blanket or anything to insulate the cold sides of the tank from the moist air (Fig. 12).

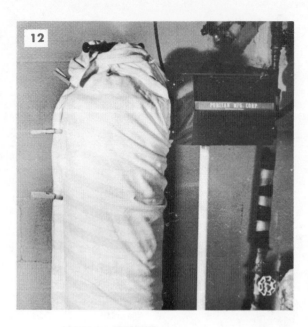

WELL PUMP

If the house is using well water, a pump and a pressure-storage water tank must be used. There are two basic kinds of pumps used in homes: the jet pump and the submerged pump. The submerged type is outside and down in the well several tens of feet. Little service can be performed on it short of replacement. The jet pump is usually located in a basement or other accessible place (Fig. 13). There is almost no simple maintenance that can be performed on it either. All bearings are sealed and no lubricating is required. Both types of pumps have a pressure switch (Fig. 14). To adjust the low-pressure cut-in and the high-pressure cut-off, see Fig. 15. To raise or lower the pressure at which the pump will start, adjust the cut-in screw. To raise or lower the pump maximum pressure setting adjust the cut-out nut. Before any adjustment is made, check the manufacturer's specifications for the maximum pump operating pressure. Any maximum cut-out switch adjustment above this pressure will make the pump run continuously never reaching an out-of-range setting. If water pressure remains low at a faucet, clean the

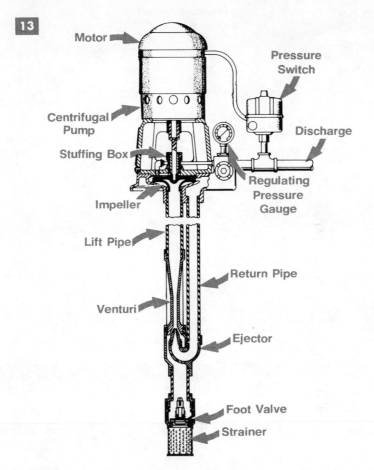

13

Motor

Pressure
Switch

Centrifugal
Pump

Discharge

Stuffing Box

Regulating
Pressure
Gauge

Impeller

Lift Pipe

Return Pipe

Venturi

Ejector

Foot Valve

Strainer

screen strainer in the aerator located on the end of faucet spout.

All pressure tanks must have an air cushion in the top to operate properly. This air cushion is maintained with an air valve (Fig. 16). If the pump cycles on and off every few seconds, the tank is water-logged and the air cushion is gone or about gone.

Air is absorbed into the water and this air cushion will completely disappear by dissolving into the water over a few weeks if it is not replaced. The air valve on the tank will normally keep a proper air cushion in the top of the tank.

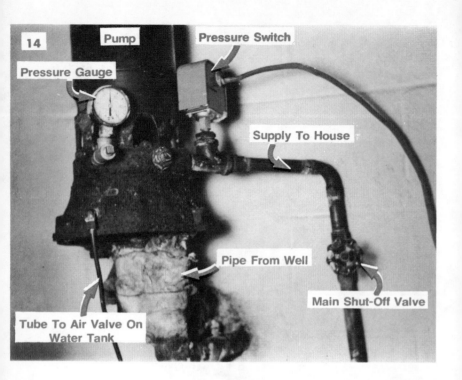

14

Pump

Pressure Switch

Pressure Gauge

Supply To House

Pipe From Well

Main Shut-Off Valve

Tube To Air Valve On Water Tank

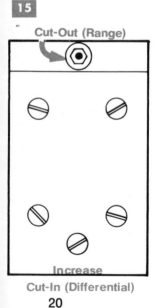

15

Cut-Out (Range)

Increase

Cut-In (Differential)

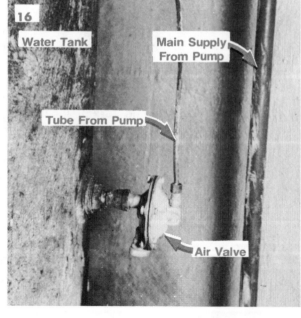

16

Water Tank

Main Supply From Pump

Tube From Pump

Air Valve

20

If it doesn't, replace it. It simply unscrews out of the side of the tank. Replacement procedure is as follows:

Step 1. Drain water from tank (shut off pump).

Step 2. Unscrew air valve assembly from tank.

Step 3. Disconnect the line to the pump.

Step 4. Buy a new valve and using pipe dope on the threads, screw into tank.

Step 5. Connect the tube from the pump.

If there is no air valve on the tank, the air cushion may be restored by:

Step 1. Shutting off the pump.

Step 2. Relieve the pressure from the tank and begin to drain it.

Step 3. Remove the plug in the top of the tank to let in air (Fig 17).

Step 4. Completely drain the tank.

Step 5. Using pipe dope, replace the plug in the top of tank.

Step 6. Close the tank-drain valve and turn on the pump.

If you have never seen the inside of a home water-pressure storage tank, shine a flashlight down into it. If you have a weak stomach, don't do it. It isn't a pretty sight because of all of the rust, scale, sediment, and other materials. Don't be alarmed to think that you drink water from that awful looking thing. What you see is perfectly harmless minerals and other materials from your well in the ground.

17

Pipe Plug

Water Closet
Flush Tank Problems

Modern plumbing codes require indoor plumbing and the use of a water-flush, water-seal toilet. The old "outhouse," unless located at a cabin or remote campsite, is rarely in use anymore for domestic human waste removal or deposit; most sanitation and health codes prohibit them. With the exception of some positive-pressure flush models, virtually all domestic water-closet toilets operate the same (Fig. 18). To be able to repair your own water-closet components you must first understand its operation.

The water that is always in the bottom of the toilet bowl is the water seal. It fills the trap portion of the bowl to prevent seepage of sewer gas into the house (Fig. 19). When the toilet is flushed, this water must be replaced. During the filling of the tank, water flows from the bowl-refill tube into the overflow tube and down through the opening into the side of the flush-valve shank on its way to the bowl (Fig. 20).

When the toilet is flushed, the trip lever is pulled. A whole series of events take place at the same time. First the tank ball pops up against the tank-ball guide and the water stored in the tank rushes down through the flush-valve seat and into the bowl. This large quantity of water carries away the waste in the bowl including the water seal. As the water in the tank lowers, the tank float rides down on the water level, opening or raising the ballcock (plunger) and its washer. As

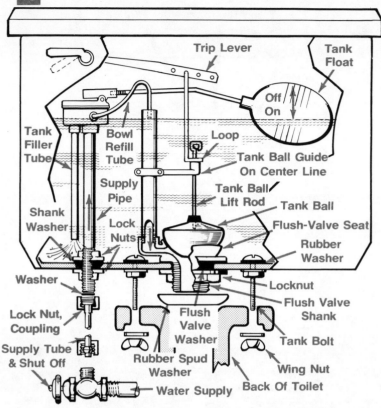

Trip Lever

Tank Float

Off / On

Tank Filler Tube

Bowl Refill Tube

Loop

Tank Ball Guide On Center Line

Supply Pipe

Tank Ball Lift Rod

Tank Ball

Shank Washer

Flush-Valve Seat

Lock Nuts

Rubber Washer

Washer

Locknut

Flush Valve Shank

Lock Nut, Coupling

Flush Valve Washer

Supply Tube & Shut Off

Tank Bolt

Rubber Spud Washer

Wing Nut

Water Supply

Back Of Toilet

the ballcock (plunger) washer lifts away from the seat in the supply pipe, water from the supply line rushes into two tubes. The tank-filler tube discharges most of the new water into the tank because it is larger. The smaller bowl-refill tube diverts a smaller amount of water into the bowl to replenish the water seal.

Meanwhile, the water level in the tank has lowered to the top rim of the flush-valve seat and the tank ball has seated firmly into the flush valve seat. As the tank begins to fill, the water becomes deeper thus exerts more pressure on top of the tank ball. This pressure helps make a tight seal to prevent water leaking from the tank. The water level slowly raises in the tank pushing the tank float higher until the (ballcock) valve plunger and its washer shuts off the flow of water from the water-supply pipe.

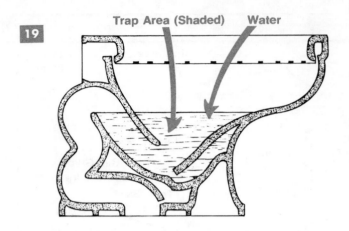

Trap Area (Shaded)   Water

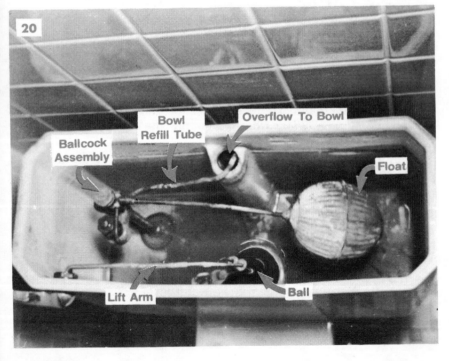

Ballcock Assembly

Bowl Refill Tube

Overflow To Bowl

Float

Lift Arm

Ball

COMMON PROBLEMS

The most common problem is one of a "running toilet." This term is applied when there is a leak between the tank ball and flush valve seat. With water continually leaking, it constantly runs in to fill the tank. The tank float never completely reaches its uppermost position to shut off the water. This is one of the main causes of high water bills that can catch you completely by surprise some month.

If a leak exists:

Step 1. Check tank-ball guide to see if it is corroded and restricting travel of the tank-ball rod (lift wire) up and down. If it is corroded, clean it.

Step 2. Check to see if the tank-ball guide is centered directly over the center of the flush-valve opening. If not carefully center it by loosening the tank-ball guide clamp on the overflow tube. Retighten clamp. The ball must be in exact vertical alignment with the flush-valve seat.

Step 3. Shut off the water supply and flush the toilet. Lift tank ball and unscrew its rod and remove ball. Check ball for worn ridges or hardened areas. Regardless of what you find, be sure to replace it; not to replace invites further trouble.

Step 4. Check the flush-valve seat for smoothness. If it is corroded or if lime deposits have built up, smooth with a 300- to 400-grit carborundum wet/dry paper dipped in water. Polish the seat smooth with the wet/dry paper wrapped around your thumb.

Another problem of water running constantly in a toilet tank is a float that is set too high. If the float is set too high, the water level raises higher than the top of the overflow tube and water runs into the bowl all the time. To stop this, simply bend the tank-float rod downward to limit the water level to one inch below the top of the overflow tube.

Check the float to see if it has any water in it. If so, replace the float by unscrewing it from the float rod. A float with water in it will not rise to the top of the water and cause the tank to fill and run into the overflow tube.

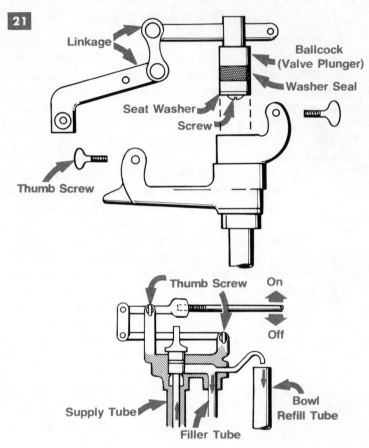

Linkage

Ballcock
(Valve Plunger)

Washer Seal

Seat Washer

Screw

Thumb Screw

Thumb Screw   On

Off

Supply Tube

Filler Tube

Bowl
Refill Tube

If water continues to flow into tank, the trouble may be a worn ballcock (valve plunger) seat washer. (Fig. 21). To remove the ballcock, remove the linkage holding it in place along with the float and float rod. Removing the two thumb-screws will allow the linkage and ballcock to be lifted out.

Ballcocks (plunger valves) come in several varieties, but all work basically the same. They have a rubber seat washer that seals against a seat. Once you understand how they work, if yours is different you can repair it also.

Seat washers are held in place a number of ways. Two common ways are by a screw or a ring nut (Fig. 22). If a ring nut is to be removed, do it carefully and don't deform it or make burrs on the outside. If pliers are necessary, put tape around the ring nut to protect it. Take the old washer to the store and get an exact replacement. If none can be found, flip the washer over and use the back side if it is smooth.

Check the seat that the seat-washer pushes against, to see if it is smooth. If not, it can be smoothed by using a 300-grit emery cloth.

The seal washers prevent water from flowing up around the ballcock and spraying water up against the top of the

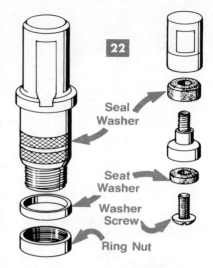

tank lid and leaking onto the floor. Some are split ring; some are not. Replace them. Install the ballcock into the toilet.

Most ballcocks have complete repair kits available at the plumbing supply store. See Fig. 23 for typical ballcock assembly parts. If in doubt, take your ballcock to the store and get the proper kit for it.

If the entire ballcock assembly needs replacement, there are several kinds to choose from. You can try to replace the old one with another just like it. However, the plastic ballcock assembly is a good one and is generally cheaper.

To replace the entire ballcock assembly (Figs. 21 and 22):

Step 1. Shut off water at the supply valve, flush the toilet, and sponge out any water left in the tank.

Step 2. Remove the locknut (coupling nut) on the water-supply tube from the ballcock shank.

Step 3. Remove the locknut from the outside on the ballcock shank under the tank.

Step 4. Lift out the entire assembly including float.

Step 5. Insert replacement and tighten outside locknut just enough to make sure that the shank washer inside the tank makes a watertight seal. Don't overtighten or the tank might crack.

Step 6. Replace the water supply tube. Test the toilet.

27

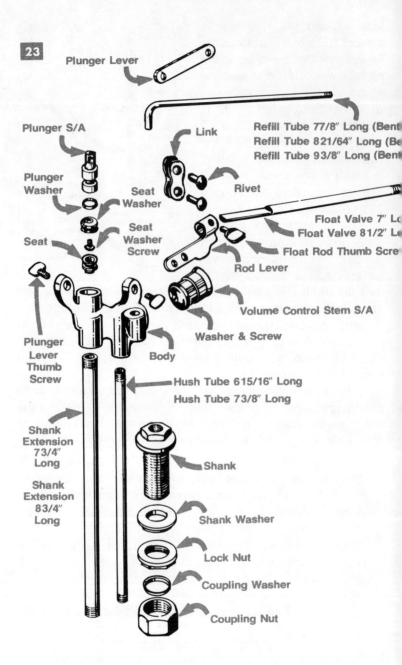

23

Plunger Lever

Plunger S/A

Link

Refill Tube 7 7/8" Long (Bent
Refill Tube 8 21/64" Long (Be
Refill Tube 9 3/8" Long (Bent

Plunger
Washer

Seat
Washer

Rivet

Seat
Washer
Screw

Float Valve 7" Lo
Float Valve 8 1/2" Le

Seat

Float Rod Thumb Scre

Rod Lever

Plunger
Lever
Thumb
Screw

Volume Control Stem S/A

Washer & Screw

Body

Hush Tube 6 15/16" Long
Hush Tube 7 3/8" Long

Shank
Extension
7 3/4"
Long

Shank

Shank
Extension
8 3/4"
Long

Shank Washer

Lock Nut

Coupling Washer

Coupling Nut

28

LOOSE TOILET BOWL AND LEAKS

Through the years for various reasons like normal use, rotted or decayed flooring, or just plain poor installation, the toilet bowl may loosen up. The way to detect this is to get down and look close where the base meets the floor and try to rock the bowl from side-to-side then from front-to-back. Any movement up and away from the floor indicates looseness than can often be tightened up easily.

Most toilet bolts through the base are hidden beneath china caps. These caps are held in place with plaster or caulking. Remove these caps by prying them up with a thin blade screwdriver or kitchen knife. Tap the caps lightly with a hammer then carefully try to drive the small screwdriver or knife under the cap (Fig. 24). Alternate tapping the cap and knife handle. Once the caps have been removed, apply some penetrating oil to the bolt threads. If you have two or four bolts through the base of the toilet bowl, tighten them all one-half turn and check for looseness again. Repeat this two or three times if necessary. Do not tighten too tight or the base will crack or break.

In some older installations, bowl rings or seals (Fig. 25) were made of caulking compound that became hard. Also some waxes become hard. This often led to leaks around the base of the bowl and the floor. If there is a leak or the bowl cannot be tightened, remove the toilet from the floor and install a new bowl ring as follows.

24

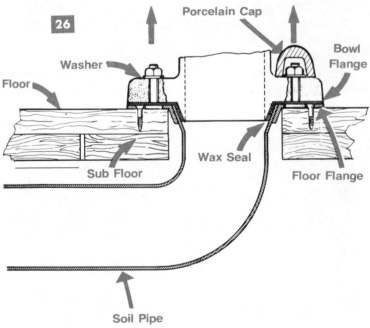

Porcelain Cap

Bowl Flange

Washer

Floor

Sub Floor

Wax Seal

Floor Flange

Soil Pipe

Step 1. Shut off water and flush.

Step 2. With a towel or rag soak all water out of the tank.

Step 3. Disconnect the water line at the base of the tank.

Step 4. Remove the nuts or bolts at base of bowl. (Fig. 26).

Step 5. Lift bowl and tank straight up if it is a one-piece toilet or close-mounted tank and toilet; if it is a two-piece tank and bowl, disconnect the water-closet or tank elbow (Fig. 27) and lift straight up.

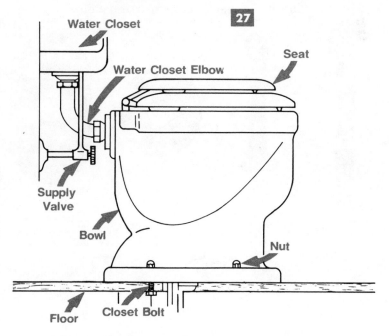

Step 6. Carefully set the bowl on its side, on a rug to prevent scratches to a tile floor or to the bowl.

Step 7. Scrape away all of the old ring or seal from base of the bowl and the floor flange.

Step 8. Place a new ring around the bowl horn and press into place (Fig. 28).

Step 9. Carefully lower the bowl vertically back over the floor flange and bolts (Fig. 29). If the wax ring has a vinyl collar, make sure it goes down into the center of the flange.

Step 10. Press down on the upper rim of the bowl and slightly twist back and forth to smash the wax ring to make a good seal (Fig. 30).

Step 11. Place washers and nuts on bolts and gently tighten the nuts. Apply caulking or setting compound inside the porcelain caps and place them over each bolt and nut.

Step 12. If it is a two-piece tank (water closet and bowl), connect the water-closet elbow.

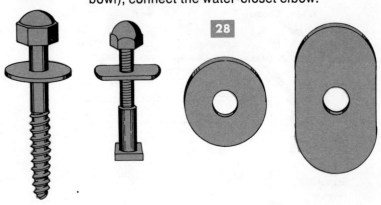

28

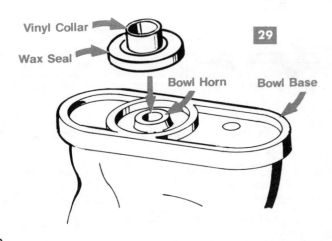

Vinyl Collar

Wax Seal

Bowl Horn   Bowl Base

29

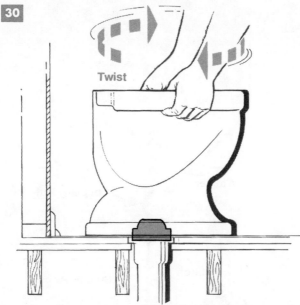

Twist

Step 13. Connect the water line to the tank. Make sure the washer at the base of the tank ballcock asembly fits up with the water line and does not leak (Figs. 31 and 32). If it leaks, replace it. Do not rotate the ballcock assembly from the centered position in the tank.

If the toilet bowl rests on the flange, the bowl must be shimmed-up off the floor to prevent this. No part of the bowl base or horn should touch or rest on the flange. To shim, raise the bowl off the floor and insert small pieces of wood near the bolts to raise the bowl base off the floor. Apply a bead of plaster of Paris or other hard-setting cement all around the rim of the bowl base. Repeat steps 9 through 13 and do not use until plaster or cement sets for 24 hours.

On a one-piece or close-connected toilet, this will raise the ball-cock connection up away from the water-supply pipe the distance the bowl was raised. There is usually enough slack in the pipe under the floor to raise up for the connection. This is only if the water pipe comes vertically out of the floor. Most newer installations have the shutoff valve coming out of the wall and the water pipe going to

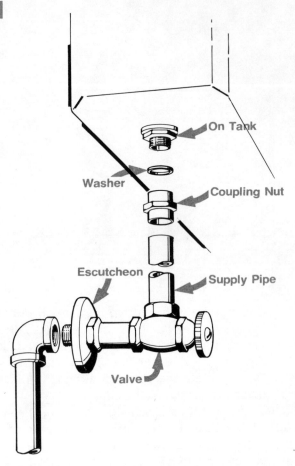

On Tank

Washer

Coupling Nut

Escutcheon

Supply Pipe

Valve

the tank from the valve is bent in the shape of a lazy-S to fit up to the base of the ballcock assembly on the outside of the tank (Fig. 32). To gain the extra distance, try straightening out the S to reach the connections. If it won't reach or fit, take the pipe to a hardware store and replace it. Bend it in a similar shape and cut length to fit. Be careful; this thin-wall pipe will kink easily.

Where a close-connected tank and bowl (Fig. 33) are encountered, they are sealed by a gasket between them (Fig. 34). If this joint becomes loose or leaks between the tank and bowl, do the following:

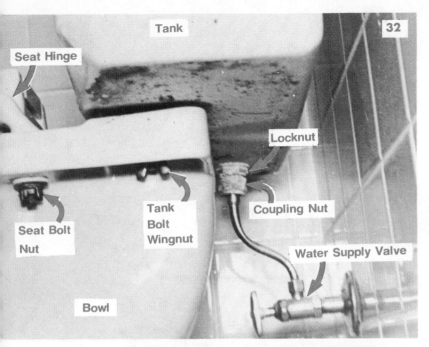

Tank

32

Seat Hinge

Locknut

Tank
Bolt
Wingnut

Coupling Nut

Seat Bolt
Nut

Water Supply Valve

Bowl

33

Step 1. Remove the nut from the tank mounting bolts and remove the tank from the bowl (Fig. 34).

Step 2. Replace the rubber gasket.

Step 3. Set the tank back onto the bowl and align the bolts carefully in the bowl holes.

Step 4. Replace the washers and nuts on the tank-mounting bolts and tighten for a snug fit. Do not overtighten.

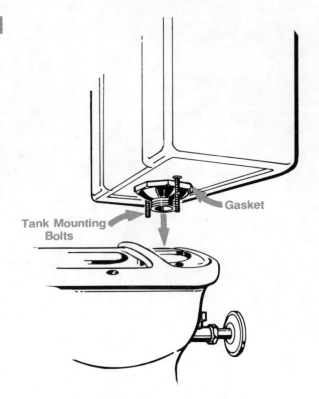

Tank Mounting
Bolts

Gasket

TOILET SEATS

If the seat lid does not set level on the bowl or touches the bowl directly, the seat snubbers are missing or worn and need to be replaced (Fig. 35). Replace the snubbers with the same kind that are removed. They may be the screw-on or nail-on type. The seat washer or seat is replaced by removing the nut from both bolts and lifting the seat out.

To buy a new seat, either take the old one with you and pick out a new one the same size or measure the distance between the mounting bolts and get a seat that mounts the same.

Don't tighten the mounting nuts too tight. Just snug the nut up against the washer that pushes against the seat rubber washer. There is no reason to make the nuts extremely tight.

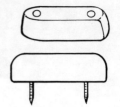
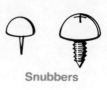

Snubbers

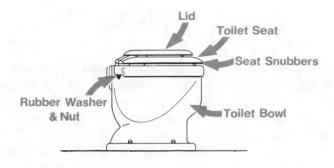

Lid

Toilet Seat

Seat Snubbers

Rubber Washer
& Nut

Toilet Bowl

Rubber Washer

SECTION 4

Septic Systems

If you live in the city or metropolitan area where your house is connected to a municipal sewer system, there is little need for you to read this section. However, if you have a septic tank or would like to gain a little insight into the underground world of sewage treatment, read on.

Septic tanks have been around for centuries in one form or another. It is an effective, although not too efficient, method of disposing of human waste. It is a system that depends upon bacteria to decompose solids and to distribute large volumes of water. The sad part about this system is the use of fresh potable (drinking) water to make it function. To a purebred plumber or anyone else who is concerned with our national resources, the thought of four ounces of urine in a toilet bowl being flushed and carried away by five gallons of fresh drinking water is a tragic waste. There are other methods of treating waste and sewage that are much more effective and efficient. They are mechanical means often used by industry and large institutions that have no municipal sewers. For the domestic home these treatment plants are presently too costly and out of reach. They also require regular maintenance and servicing where a septic tank does not. So until something better comes along, learn to take care of your septic system.

If something goes wrong with your septic system, there is little you can do except call a firm to come and pump out the tank (Fig. 36). If you have taken care of your septic tank system, the tank may not need pumping within ten

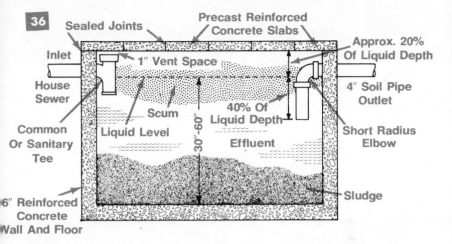

years. However there are several things that may require it to be pumped more often and are mentioned in the discussion that follows.

The percolation of the soil around a house determines how well the finger system or water-disposal field does its job. Some mountainous, rocky, or shale, ground areas do not carry away the water efficiently, even when only small amounts of water are put into the system. The size of a finger system is determined by the size of a family dwelling, percolation of the soil, and how much water will be expected to be disposed of (Fig. 37). However, many houses in some areas have septic systems that must dispose of much more water than just human waste and waste from the kitchen sink.

Wash and sump pump water is often discharged into the septic system. There is very little reason for this except to dispose of the huge amounts of water that these appliances handle. Most systems handle it well—except in very wet or rainy seasons. The first symptom of a water logged disposal field or finger system is that the toilet won't flush at all or is slow in clearing water from the toilet bowl. The septic system is to be suspected only after everything has been checked out as listed in this book under clogged drains and toilets. If the septic system is sluggish to handle all of the water, first cut down on the amount of water put into it. If the wash water

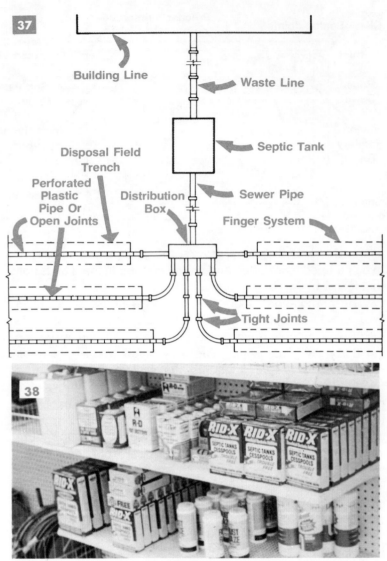

37

Building Line

Waste Line

Disposal Field Trench

Septic Tank

Perforated Plastic Pipe Or Open Joints

Distribution Box

Sewer Pipe

Finger System

Tight Joints

38

goes into it, go to the laundromat for several days. Add a package of concentrated bacteria additive (Fig. 38) by flushing it down the toilet. If the sump pump goes to the septic system, temporarily run plastic pipe or hose from the sump pump out into the yard for several days. Of course, have the tank pumped out if all else fails.

Since a septic system that is troublesome or does not work is much more than an inconvenience, begin to think "plumbing care" about your septic system. When septic tanks are sluggish and slow to carry away the water when the toilet is flushed, solids and paper begin to settle and accumulate in the waste or sewer lines from the toilet to the septic tank. When this happens often enough, there is double trouble. Not only is the system not getting rid of the water but these waste lines are becoming clogged.

First of all give serious thought to what you can do to reduce the amount of unnecessary water going into the system.

First on the list is to get the sump water out of it. Most sumps contain relatively clean water. Most often with no sewers, the house foundation drain tiles empty into the sump. This water is not sewage. Pump this water into the yard or a ditch at least during wet seasons.

Wash water contains various kinds of soap. Most widely known are the multitude of detergents. Regardless of what you think of the ecology movement—for or against it—detergents do not completely decompose. If the wash water must be pumped into the septic system then at least use a non-phosphate or biodegradable laundry soap. It doesn't matter where the wash water is disposed; the soap in it should be biodegradable.

Do not let any petroleum products, paint thinners, solvents, cleaning fluids, cigarettes, or any items of this nature get into the septic system. Most of these things are harmful to the bacteria. The bacteria that grows, feeds and breeds in the septic tank decomposes the solids into sludge that settles to the bottom and other lighter-than-water items that float to the top and form scum. This bacteria is the first line of defense in treating and purifying the raw sewage. It doesn't do all that good of a job even in the best operating systems. A good percentage of raw sewage goes on through the finger system and into the disposal field. The natural bacteria found in the ground from two to six feet deep also decompose and purify some of the sewage water that was not treated in the tank. The remaining percentage of untreated sewage is filtered into the ground and is generally, or hopefully, purified before the water finds it way back into a fresh water source like your own well. So take care of the bacteria, it does an important job.

Waste from the kitchen sink invariably goes to the septic tank. Since it contains grease, animal fats, and food particles, it cannot be pumped on the ground surface. Grease in the septic tank and sewer or waste lines going to it isn't all that harmful. It floats and gathers with the soap and other items in the septic tank to form the scum at the top of the tank. When the tank is pumped out, most of this is removed. Proper care and maintenance dictates that the amount of grease going down the kitchen sink drain be reduced as much as possible. Pour all excess cooking grease and fats into the trash. Wipe and carefully scrape all pots, pans, and dishes clean before immersing them in the soapy sink water. Water from a brine charging water softener is not all that good for a septic tank either.

Garbage disposal units do effectively operate in many septic tank systems. The septic tank bacteria does a fair job of attacking and helping decompose food particles that reach the septic tank. Obviously a garbage disposal puts large portions of solids in a septic tank and this adds greatly to the accumulation of sludge at the bottom and scum at the top. The incidence of septic tank cleaning in this situation is higher than a comparable house without one. If you don't have a garbage disposal unit in operation going into your septic tank, don't get one regardless of the amount of flap you get from your wife.

Other items that add to septic tank solids accumulation are the "modern" flushable sanitary napkins. As recommended under the section on toilet problems, don't allow these to be flushed, if at all possible. Of course, the regular sanitary napkins should never be flushed down the toilet.

Trees can cause septic system problems. Small water seeking roots from any tree will proliferate in a finger system or disposal field. Several trees near and over the septic system will cause many problems when those trees become very large. A common blockage in the sewer or waste line from the house to the septic tank is roots that find their way into the pipe and restrict the flow of solids and paper. A solid blockage soon occurs. If you have a root problem, a tree or two may have to be removed, or have cast iron soil pipe installed.

Large trucks, especially concrete trucks will crush and collapse any septic tank. They might damage good finger systems. A strict rule to follow is don't let any truck or vehicle drive or back across the area where the septic system is installed, especially the septic tank. Take care of the septic system and it shouldn't cause any problems for many years.

SECTION
5

Water Heaters

Water heaters often give years of trouble-free service. There are a couple of preventive maintenance jobs you can do to extend heater life.

One of the things to prolong the life of any water heater has already been covered in this book under water conditioning equipment; use conditioned, rust-free water. If the water in the household is well water and it is known to be "rusty" or extremely "hard" then this same suspended rust and "lime" is building on the inside of the water heater. This is in addition to the normal accumulation of scale and deposits that form inside a heated tank. Get a water softener and/or iron filter.

Most modern water heaters are so-called "glass lined." This is a nonmetallic coating of glass or ceramic that prevents the water inside the tank from contacting the metal sides and producing rust and scale. Regardless of what precautions are taken, some scale will still form in a tank along with small amounts of harmless dirt and sediment. That which is not attached to the tank walls will settle to the bottom. The second important preventive maintenance job to be performed is to drain three or four gallons of water from the hot water heater tank every six months. This is done by opening the drain faucet on the side of the tank (Fig. 39). This is most important for an efficient gas heater because the heat flame is at the bottom of the tank.

Gas water heaters are very effective and certainly are the less expensive to operate when the cost of operation is

44

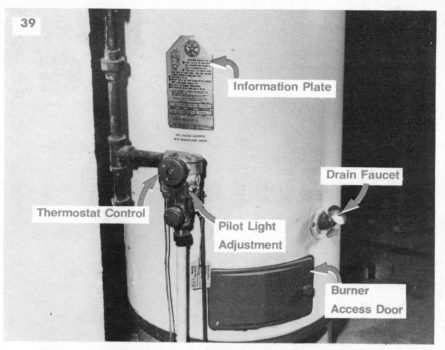

39

Information Plate

Drain Faucet

Thermostat Control

Pilot Light
Adjustment

Burner
Access Door

compared to electric water heaters. This consumer-conscious value, however, doesn't come without any strings attached. Gas heaters must be well vented when installed. Anyone who has a gas water heater in the basement, garage, or utility room must always remember one very important fact. Any gas appliance has an open-flame pilot light. This is not much larger than the flame given off by an old-fashioned kitchen match and is quite safe. However, flammable liquids should never be used anywhere in the same room or floor level where gas appliances are located.

All water heaters have a thermostat control on the side to regulate the water temperature (Fig. 39). Adjust it slightly upward or downward to suit the hot water needs of the house. Rarely do domestic hot water temperatures need to be above 160 degrees. Adjust to suit your needs between 140°F and 160°F.

Pilot lights do go out on gas water heaters. They can be relighted by anyone except a small child simply by following the instructions on the information plate.

Washing Machines

Washing machines generally have only one kind of plumbing problem—water won't enter the tub. First check to see if the valve or faucets supplying the tub are open. If all of the other faucets in the house have enough pressure, shut off the washing machine hot- and cold-water faucets and unscrew the supply hoses from these faucets (Fig. 40). Turn on the faucets and check for adequate pressure. If water pressure is low or missing from any one or both faucets, check for a bad or stuck seat washer on the faucet spindle. Service or replace as covered in the section on faucet re-

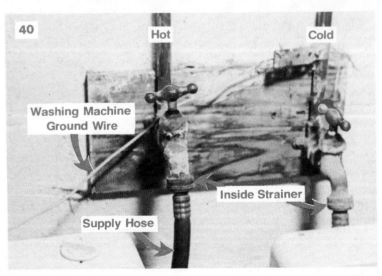

40   Hot   Cold

Washing Machine
Ground Wire

Inside Strainer

Supply Hose

pairs. Check the hose strainer to see if it is clogged. If corroded or clogged, clean or replace it. These can be bought at most hardware and home appliance department stores. Remove both water supply hoses from the back of the washer. Often a strainer can be found here. Clean or replace it. Once all of these things have been checked, connect them all back up again using new hose washers and try the machine. If water still will not enter the machine, then the solenoid valve in the washer or the controls for this valve are at fault. Such repairs are beyond the scope of this book and should be covered in most home appliance repair books.

All washers have a water discharge hose. For the machine to function properly, this hose needs to be almost as high as the top of the machine. If left to dangle low or on the floor, the tub water can drain out. The wash water must be pumped out by the internal pump up and over the highest part of the hose. The end of this hose should have a lint strainer if the wash water is discharged into a sum pump (Fig. 41).

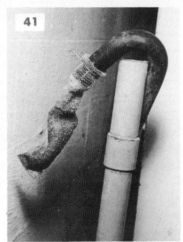

Sump Pumps

Sump pumps are designed to remove water from sumps and other areas where water collects. A sump is the pit that collects the water in quantity for removal by pumping (Fig. 42). A sump should not be so small that water raises and lowers rapidly when filling or pumping. It should be large enough to catch water during peak water periods so that the pump won't cycle on and off any sooner than three to five minutes—preferably longer. Pumps can run continuously for long periods of time, weeks and months if necessary. Constant cycling on-and-off every few seconds or minutes should be avoided for long motor life.

If you have a water problem that collects in a crawl space or basement, install a sump and a pump. Almost all basements and crawl spaces have adequate sand and gravel backfill. This is usually suitable for installing a sump pit. A good inexpensive sump pit can be made from one piece of vitrified tile 20″ in diameter and 18 to to 24″ long. If under a crawl space, a hole can be dug (with much agony) and dirt filled in around the tile placed vertically in the ground. If in a basement a hole will need to be cut through the concrete floor and then a hole dug. In either case do not seal or close up the bottom. Leave it open to the gravel.

Buy a suitable sump pump made for home use that has a 1½ or 1¼″ discharge outlet. Plan where to discharge the water either in a sewer, dry well, septic tank, or outside ditch. Buy enough plastic pipe to reach the distance and connect

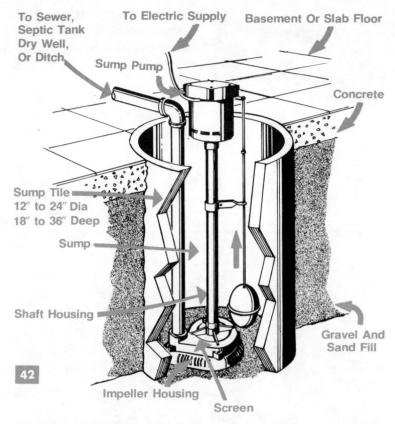

To Sewer, Septic Tank Dry Well, Or Ditch

To Electric Supply

Basement Or Slab Floor

Sump Pump

Concrete

Sump Tile 12″ to 24″ Dia 18″ to 36″ Deep

Sump

Shaft Housing

Gravel And Sand Fill

42

Impeller Housing

Screen

it to the pump discharge. A typical plastic pipe connection using clamps is shown in Fig. 43.

The electrical outlet for the pump should be fused with a "slow-blow" type of fuse because of high current drain on motor initial start-up. Adjust the float switch so that the pump will remove at least 8 to 10 inches of water from the sump at a time.

Whether a submersible or standup pump is used makes little difference. Submersible pumps are more costly but may be necessary in a crawl space where vertical space is limited. The "stand-up" type is less expensive and does a good job. For a few dollars more, a pump can be bought with a brass impeller housing and a brass impeller shaft housing. They are worth it because water doesn't rust or

corrode them. Many new pumps now have durable self-cleaning plastic impellers that are long-lasting and trouble free.

If you install a pump or already have one, there are a few things to consider. Many house foundation drain tile discharges the water into a sump. Washing machines frequently discharge the water into a sump. Water seepage

from basement walls and up through basement floor cracks eventually finds it way into the sump. Charging water from a brine water softener is often discharged into a sump. None of this water is dangerous hygienically, nor is it sewage.

To keep sump pumps working properly, the strainers must be cleaned monthly (Fig. 44 A and B). Those that have self cleaning impellers and no strainers require little maintenance. If wash water is discharged into the sump, a strainer to catch the lint should be used. (Fig. 45). This will prevent large and rapid accumulations of lint on the pump strainer screens.

If at all possible, sump and wash water should not be sent to a septic tank for the reasons discussed in the section on septic systems. However, if there is no other choice, millions of households are operating that way.

In areas where heavy water accumulation during rainy seasons could pose a threat to valuable possessions if the pump fails, use another pump as a back up (Fig. 46). Don't take chances. If the pump fails, it will do so during the rainy season. Several inches can accumulate in a basement in a few hours during continued heavy rains.

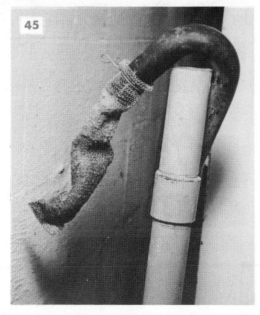

Float Switch

Water Softener
Discharge Line

Drain From
Wash Tub

Submerged
Pump

Float

If a secondary or back-up pump is installed, its piping and water discharge should be separate from the primary pump. The quickest and simplest method is to pipe the flow out a basement window or through a wall into the yard. The electrical wiring should be on a circuit different than the primary pump. If a motor seizes and blows a fuse, both pumps would be dead if they were on the same fused circuit.

Miscellaneous Plumbing Repairs

Water tanks sometimes develop leaks after they have been in use for 10 to 15 years. How soon simply depends on the quality of the tank and the water. Some tanks never develop a leak.

If your water tank develops a leak it will usually start out as a seep with a small puddle on the floor. More severe pinholes will put out a little stream. Quick but temporary repairs can be done with a self-tapping plug (Fig. 47). This plug has a small washer that helps stop the leak.

Get several sizes, because the holding strength around a rusted hole is not great. Too small a self-tapping plug won't hold or will strip out when tightened. If this happens, use a larger size.

Drain the tank. Gently tap the tip of the plug into the pin hole then screw the plug into the hole. Make sure the surface around the hole is smooth for the washer to make a good seal. In severe cases, an epoxy or tub caulk can be spread around the plug body before screwing in. If this is done, let set for 24 hours before use.

This repair is only temporary and should provide you with enough lead time to get a new tank.

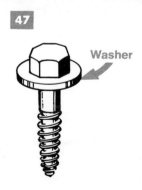

47

Washer

SOIL-PIPE JOINT LEAKS

Cast-iron soil pipe is a bell and spigot joint joined and sealed by a packing called "oakum" and hot lead (Fig. 48). If for various reasons this joint begins to leak, water will seep around between the lead and wall of the pipe.

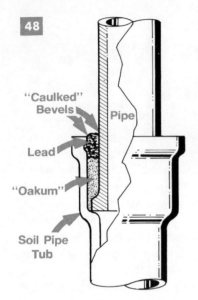

48

"Caulked" Bevels

Pipe

Lead

"Oakum"

Soil Pipe Tub

Such a joint should not leak because the soft lead is carefully "packed" by a beveled caulking iron after the molten lead hardens in the joint. (Fig. 48). Sometimes such leaks can be stopped by light tapping with a large broad-blade screwdriver and a hammer (Fig. 49). Do not drive the lead down hard into the joint. The purpose of lightly tapping all around the circumference of the joint is to produce two small bevels as shown in Fig. 48 which will "tighten up the joint."

If the joint still leaks after trying to "recaulk it," use

49

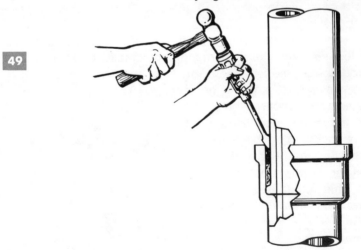

epoxy glue. If possible, do not run water through the pipe for 24 hours. Clean all around the joint with a wire brush, then apply a properly mixed layer of epoxy cement or glue around the joint. Let set for at least 24 hours. In the place of an epoxy glue, a liquid or plastic steel can be used.

SCREWED-PIPE LEAKS

Leaks around fittings or valves that have been joined by threads can often be stopped by one of four methods. First try to tighten the joint with a wrench by turning that fitting or pipe with a wrench in a clockwise direction. Always use another wrench (backup) to hold the fitting or pipe from turning when the tightening action is applied. This remedy often causes other problems and a joint is loosened somewhere else. So unless other joints are considered, this method is the least satisfactory.

Using method two, shut off water and drain all water from the pipe by opening a faucet someplace in the house lower than the leaky joint. Carefully dry the joint and heat it lightly with a torch or heat lamp. Do not get it hot, just good and warm to the touch. Clean the joint with a wire brush and steel wool. If necessary use a rust remover, dry, and then polish all metal surfaces bright and clean. Apply epoxy cement or liquid plastic steel and let harden for 24 hours.

Method three employs the use of a plastic pipe dope that comes in a roll-like tape. If all else has failed, disconnect the joint by unscrewing the pipe from the fitting. Clean and remove all previous hardened and cracked pipe dope from the male threads of the pipe and female threads of the fitting. Apply one complete wrap of the tape dope around the pipe threads (Fig. 50). Screw the pipe back into the fitting using one wrench on the pipe and one wrench on the fitting. When turning the pipe or the fitting make sure the other one is held by the backup wrench and doesn't turn. If the leak still persists and you have no knowledge about threading, cutting, and fabricating screwed pipe, method four is to call a plumber or take a plumbing course in a local night school.

COPPER JOINT LEAKS

If a leak occurs around a copper joint, resoldering that joint often will stop it. Shut off the water in the pipe and drain all the water from it. It is important that all water be drained away from the leaky joint or else the temperature can never be raised above 212° F to melt the solder.

Carefully dry the joint and clean all around the joint with fine emery cloth or steel wool (Fig. 51). Apply a coat of

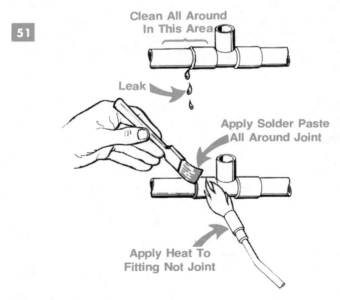

Clean All Around
In This Area

Leak

Apply Solder Paste
All Around Joint

Apply Heat To
Fitting Not Joint

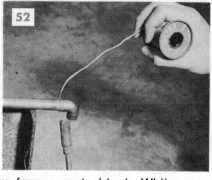

solder paste all around the joint and apply heat with a small torch. The heat must be applied to the fitting not to the solder paste. When the proper temperature has been reached, the solder will be sucked into the joint (Fig. 52) and the paste will change colors from grey to black. While the joint is still hot, take a damp cloth and wipe lightly around it.

Use caution when soldering in closed places or near flammable materials. Make sure the torch does not ignite anything.

To solder a new or old copper joint that is apart, you need the equipment shown in Fig. 53.

1. Clean end of pipe with emery cloth to a bright and shiny appearance (Fig. 51).
2. Clean the inside of the fitting (elbow, tee, etc.) with emery cloth.

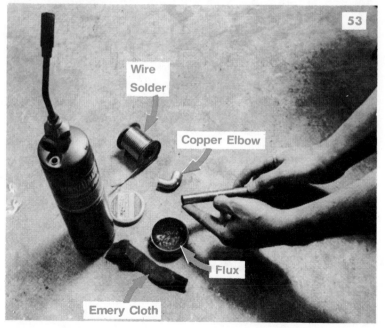

Wire Solder

Copper Elbow

Flux

Emery Cloth

3. Apply soldering flux to both the pipe and fitting (Fig. 51).
4. Insert the pipe into the fitting.
5. Apply heat to the fitting from underneath. Let the heat travel upward (Fig. 52).
6. Apply solder at the top of the joint and let it run into the joint.
7. Remove heat and wipe with a cloth.

JOINING PLASTIC PIPE

Most do-it-yourself plumbers need an understanding of joining plastic pipe. Any plumbing piping job where new pipes are run or old pipes are sometimes replaced, can be done with plastic pipe. Plastic pipe is safe, easy to install and inexpensive.

All plastic pipe can be cut with a hacksaw, or any saw that will cut wood. It can be fabricated to do and fit almost any job. There are two methods to be concerned about when joining plastic pipe.

The simplest method of joining plastic pipe is using fittings (Tees, elbows, adapters, etc.) that insert into the pipe and are held in place by stainless-steel clamps around the pipe (Fig. 54). These joints need no solvent, dope, or cleaning preparation. When assembled as shown, they will not leak and give long service (Fig. 43).

The other method is for joining PVC (polyvinyl chloride) pipe. This pipe is rigid and usually does not come coiled. It slips into the fittings much the same as copper pipe. It can be cut easily with almost any saw. The joints are coated inside the fitting and outside the pipe with a solvent or

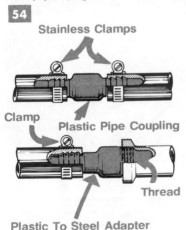

54

Stainless Clamps

Clamp

Plastic Pipe Coupling

Thread

Plastic To Steel Adapter

glue. (Fig. 55). Once the pipe is cut to proper length and the solvent applied, it is inserted into the fitting with a slight twisting motion to distribute the solvent. After a few hours, the joint and pipe are ready for use.

SWEATING TANKS AND PIPES

Toilet, well pump, and other water tanks will invariably sweat anytime the humidity is high, the air is warm and the surfaces of the tank are cooler than the surrounding air. Cold-water pipes will do the same. The scientific principle

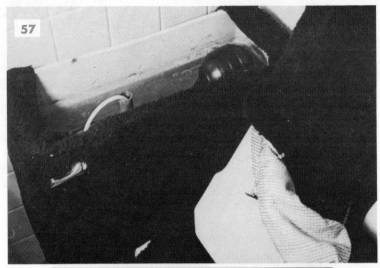

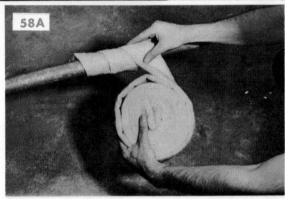

involved to prevent sweating is to insulate the cold water pipes. Insulate the cold water pipes? Certainly! The purpose is to keep the warm moist air from coming in contact with the cooler pipe or tank surface.

Insulation for water tanks can be a simple old blanket no longer used on the bed (Fig. 56). Wrap it around the tank and fasten with clothes pins. Toilet tanks should be insulated with a tank and lid cover sold as a bathroom set in most department stores (Fig. 57). Pipes can be wrapped with anything from rags taped in place to cover the pipes to roll insulation bought in a hardware store (Fig. 58).

CRACKED BATHROOM CHINA

If a toilet tank or bowl should develop a crack for any reason, it should be carefully observed. If the crack is severe enough, the fixture may need to be replaced. If the crack isn't a structural failure that threatens to collapse the fixture, then it probably can be sealed. A tiny crack can quite effectively be sealed, if it can be reached, by using any modern epoxy cement or bathtub caulk. Cleaning the surface is most important and the use of a lacquer thinner or a safety solvent that replaced carbon tetrachloride should be used. Apply the sealer or cement as directed on the can or tube. If a leak persists, or develops, then replace the toilet bowl or tank or both. Try a different color this time.

FROZEN PIPES

Many house trailer dwellers have experienced frozen pipes of one kind or another. The most common is water supply lines but drain lines also freeze. Regardless of where frozen pipes occur, under the trailer, in the garage, in the unheated crawl space, etc., there is one easy cure. Electric heating tapes or cables for preventing and thawing frozen pipes cost but a few dollars and do the job well.

Buy the length of cable that will let you wrap the entire length of pipe to be protected. They can be bought with

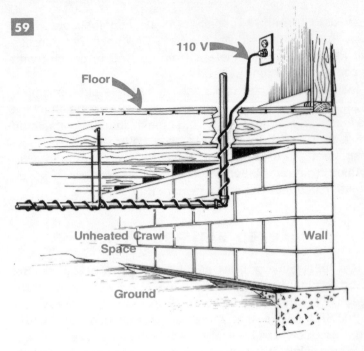

110 V

Floor

Unheated Crawl Space

Wall

Ground

thermostats or without. Follow the instructions provided. Many heat-wrap cables can be left on indefinitely and in severe weather some can be covered with insulation for further protection (Fig. 59).

A small butane torch can be used to thaw frozen pipes. It does an effective job if the flame is played along the entire length of the frozen section. Remember that thawing frozen pipes with any amount of heat takes time and patience.

SECTION

Plumbing Supplies

Many persons who attempt to repair anything, if they are not familiar with the job will disassemble or tear it apart and then determine what parts are needed. This is not un-common for do-it-yourself plumbing repairs since most people know very little about plumbing tools, parts, equip-ment, and supplies. How do you go about learning the supplies, terminology, and people associated with plumbing? It isn't too difficult. Most journeyman plumbers (union ones at least) must go through a five-year apprenticeship training program. They virtually know it all. There is no reason for you to know that much but there are a few things you should be aware of and do.

Regardless of where you go, you must know what you need and be able to call it by a name that makes sense to a salesperson.

Try to determine the name of the manufacturer who made what you need. Get a model name or style name or number. If nothing else, measure dimensions of important things like distances of hole centers, locations of holes, heights, etc. Make a sketch of it or even take a polaroid picture of it and take that along. The best method, if possible, is to take the actual defective part or fixture with you when you go after the plumbing supplies. Before you go, try to make a detailed list of everything needed. Even then several things are often forgotten.

The salesperson that you talk to may or may not be able to give much help. Of course, if you have worked with plumbing much you will be at ease in even the largest supply house and be able to make your needs known. If you are like everyone else you will want to ask a lot of questions and cling to every word of advice or suggestions that is mentioned by the salesperson or counterman.

First of all, regardless of where you go, lay it on the line and ask for help. Explain your problem as clear and precisely as possible. Don't try to impress anybody with what you think or wish you knew.

The most help can be obtained by going to a plumbing supply house or local plumbing shop. Some large plumbing supply houses sell to mechanical contractors and plumbing shops and might not appreciate your small business. But try them anyway. Many of them genuinely want it. Some plumbing shops won't take the time with you either. However there are some shops that like to sell to the do-it-yourself plumbing trade. Try them all. A disadvantage to going to these place is that normal business hours are often five days a week from 8:00 A.M. to 4:30 P.M. The average person can't get to them during those hours. That pretty well leaves you with the hardware and large chain department stores that have plumbing departments and who are open until 9:00 P.M.

Sometime when there is no need or emergency stop in a good hardware store and slowly browse around the plumbing departments. Much can be learned by carefully studying what they have on display. Read labels, directions and instructions. Part of your need for plumbing knowledge is just knowing what supplies are available.

When you really need plumbing supplies seek a floor salesperson and ask for help. There is little guarantee that the person who helps you will know much more about plumbing than you. But between the two of you, something can be worked out.